HANDBOOK OF INORGANIC QUALITATIVE ANALYSIS

A STEP BY STEP GUIDE FOR DETECTION OF MIXTURES OF INORGANIC SALTS

RARE ELEMENTS INCLUDED

Dr. Maharudra Chakraborty (Ph.D.)

SCIFINITY PUBLICATION

HANDBOOK OF INORGANIC QUALITATIVE ANALYSIS

Copyright © Maharudra Chakraborty

All rights reserved. No part of this publication may be reproduced, distributed, or transmitted in any form or by any means, including photocopying, recording, or other electronic or mechanical methods, without the prior written permission of the publisher, except in the case of brief quotations embodied in critical reviews and certain other noncommercial uses permitted by copyright law.

First Edition: 2019

SCIFINITY PUBLICATION

Also available in e-book format

Preface

This book is intended for the undergraduate and post graduate students in colleges and universities. Qualitative analysis in their inorganic chemistry practical courses plays a vital role to grow some knowledge in this field. However, most of the students feel difficulties during the detection of ions in inorganic salt mixtures because some ions in that mixture of salts interfere to the typical reactions of other ions.

This book includes a systematic approach for detection of inorganic acid and basic radicals when they are mixed together, even if the mixtures include more than six ions. Most of the difficulties and their removals have been given in this book so that student can easily report all the ions in their practical exam. Moreover, some special method for detection for a few complicated radicals are also included.

For the postgraduate students, detections of some rare elements are also discussed.

Hopefully, this book can be very much helpful to students and also to teachers of degree colleges and universities globally.

I am very much thankful to my teachers for their teachings, otherwise this book could not be successful.

Dr. Maharudra Chakraborty

Dedicated to Science

CONTENTS

	CHAPTERS	PAGE
1.	**Colors of Inorganic Salts**	1
2.	**General Solubility of Inorganic Samples**	4
3.	**Scheme for Inorganic Qualitative Analysis**	6
4.	**Special Notes on Mixtures of Ions**	40
5.	**Difficulties and Their Removals During Group Analysis**	44
6.	**Addition of Some Common Reagents in Water Extract**	46
7.	**Detection of Rare Elements**	50
8.	**Bibliography**	52

1

Colors of Inorganic Salts

A. Guesses on solid samples

Color	Formula
Blue	Cu(II), Co(II)
Black	PbS, CuS, CuO, HgS, FeS, MnO_2, CoS, NiS, Ag_2S, CoO, NiO, Sb_2S_3, Sb_2S_5
Red Brown	Fe_2O_3, $PbCrO_4$, Cu_2O, PbO_2 (chocolate)
Yellow Brown	CdO, SnS, Ag_3AsO_4
Yellow	CdS, CdO, As_2S_3, SnS_2, PbI_2, HgO, PbO, $K_4Fe(CN)_6$, $MCrO_4$ (M = any metal), $FeCl_3$, $Fe(NO_3)_3$, AgI
Light Yellow	Bi_2O_3
Orange Red	$(NH_4)_2Cr_2O_7$, $K_2Cr_2O_7$, $K_3Fe(CN)_6$, Sb_2S_3
Red	HgO, Pb_3O_4, As_2S_3, HgI_2, HgS, Sb_2S_3, Cu_2O
Pink	Mn(II), Co(II), MnO_4^-, Cr(III) alum
Green	Cr_2O_3, Hg_2I_2, Cr(III), Cu(II), Ni(II)

B. Guesses on solution

Blue	Cu^{2+}
Green	Ni^{2+}, Fe^{2+}, Cr^{3+}
Yellow	CrO_4^{2-}, $Fe(CN)_6^{4-}$, Fe^{3+}
Orange Red	$Cr_2O_7^{2-}$
Purple	MnO_4^-
Pink	Co^{2+}, Mn^{2+}

N.B.: Blue HCl solution + water → pink → Co^{2+} maybe
Colorless HCl solution + water → turbid → Sb^{3+}, Bi^{3+}, Sn^{2+} maybe

C. Colors of insoluble samples

Green	Cr_2O_3
Dark Red	Fe_2O_3
Bronze	SnS_2
Dark Yellow	$PbCrO_4$
Brown	$Cu_2[Fe(CN)_6]$
Blue	$Fe_4[Fe(CN)_6]_3$
Pale Yellow	AgI
Very pale yellow	$AgBr$
White	$AgCl$, $PbSO_4$, $SrSO_4$, $BaSO_4$, CaF_2, SnO_2, Al_2O_3, SiO_2, $CaSiO_3$

D. Colors of aqua regia soluble samples

Yellow	As_2S_3
White	Hg_2Cl_2, Hg_2Br_2
Black	NiS, CoS, HgS, PbS
Red	HgS, HgI_2
Mosaic Gold	SnS_2

E. Colors of the precipitates in Group Analysis (See later chapters)

Group	Precipitates	Color
Gr-I	$PbCl_2$	white
	$AgCl$	white
	Hg_2Cl_2	white
Gr-IIA	HgS	black
	PbS	black
	CuS	black
	Bi_2S_3	Choco-late brown
	CdS	yellow
Gr-IIB	As_2S_3	yellow
	Sb_2S_3	orange
	SnS	brown

Gr-IIIA	Al(OH)$_3$	white
	Fe(OH)$_3$	Red brown
	Cr(OH)$_3$	Dirty green
Gr-IIIB	ZnS	white
	MnS	Flesh pink
	CoS	black
	NiS	black
Gr-IV	CaCO3	white
	BaCO3	white
	SrSO3	white

* Rest of the inorganic compounds are mostly white in color.

2
General Solubility of Inorganic Samples

A. Notes on water solubility of inorganic salts
- All nitrates (NO_3^-) are soluble in water except some basic nitrates such as $BiONO_3$.
- All nitrite (NO_2^-) salts are water soluble except $Ag(NO_2)$ [hot water soluble].
- All Na^+, K^+ and NH_4^+ salts are soluble in water.
- $S_2O_3^{2-}$ salts are water soluble except for Ba^{2+} and Ag^+ which are sparingly soluble.
- BO_2^- (borates) & SiO_3^{2-} salts are insoluble in water except alkali metal salts.

B. List of Insoluble inorganic samples
All the following inorganic salts are insoluble in water as well as in concentrated acids (HCl, HNO_3 etc. even in aqua regia).

White	$AgCl$, $PbSO_4$, $SrSO_4$, $BaSO_4$, CaF_2, SnO_2, Al_2O_3, SiO_2, $CaSiO_3$
Colored	Cr_2O_3 (Green) Fe_2O_3 (Dark Red) $PbCrO_4$ (Brown or Dark yellow) AgI (Pale Yellow) $AgBr$ (Very Pale Yellow)

C. List of Aqua Regia (3 :1 conc. HCl and HNO₃ mixture) soluble salts

Yellow	As_2S_3
White	Hg_2Cl_2, Hg_2Br_2
Black	NiS, CoS, HgS, PbS
Red	HgS, HgI_2
Mosaic Gold	SnS_2

- Hg_2Cl_2 and Hg_2Br_2 turn black on addition of NH_4OH. This is an indication of their presence.

3
Scheme for Inorganic Qualitative Analysis

The following abbreviations are used for the following:

dH   Dilute Hydrochloric acid

cH   Concentrated Hydrochloric acid

dN   Dilute Nitric acid

cN   Concentrated Nitric acid

dS   Dilute Sulfuric acid

cS   Concentrated Sulfuric acid

liq.   liquid

tt   test tube

soln.   solution

ppt.   precipitate

AcOH   acetic acid

MeOH   methyl alcohol

DRY TESTS FOR BASIC RADICALS

EXPERIMENT	OBSERVATION	INFERENCE
1. Ignition test Solid sample in a dry test tube, heat	1(a) Sample melts on heating.	(a). Maybe K^+, Na^+ (white liq.) $PbCl_2$ (yellow liq.)
	(b) Change of color (i) Yellow when hot, white when cold (ii) Yellow (iii) Blackening (iv) Green	(b) Maybe (i) Zn-salt ($ZnSO_4$ does not respond) (ii) Pb - salt (iii) Cu, Fe, Ni or Co-salt (iv) Cr-salt
	(c) Sublimate formation i) White sublimate, pass H_2S 1) Yellow, 2) Orange 3) No change (ii) Yellow (iii) Orange (iv) Black	(c) May be (1) As_2O_3, (2) Sb_2O_3 (3) NH_4Cl, H_3BO_3 (ii) HgI_2, As_2S_3 (iii) Sb_2S_3 (iv) Hg
	(d). Evolution of gas i). Violet gas ii). Brown gas	(d) May be (i). I- salt (I^- + Oxidizing Agent $\rightarrow I_2$) ii). NO_3^- (of Pb, Bi etc.)

	iii) Smell of ammonia	iii) NH_4^+ (NH_4^+ salt + base)
2. Soda lime test (in case of white or grey sublimate in test 1) solid sample + Na_2CO_3 or soda lime, (3 times), heat.	2(a) Smell of NH_3, turning $Hg_2(NO_3)_2$ -paper black	2(a) NH_4^+ – salt
	(b) Grey mirror giving globules on rubbing	(b) Hg – salt
	(c) Garlic odor	(c) As – salt
3. Flame test: A clean Pt-wire, moistened with cH, is touched with the solid sample and then held at the edge of oxidizing flame.	3(a) Golden yellow flame, invisible through double blue glass	3(a) Na salt
	(b) Violet (crimson red through double blue glass)	(b) K. salt
	(c) Light green	(c) Mn salt
	(d) Bluish green	(d) Cu-salt
	(e) Apple green	(e) Ba-salt
	(f) Crimson-red	(f) Sr-salt or CaF_2
	(g) Brick-red (transient)	(g) Ca-salt

	(h) Bluish white (asbestos fiber)	(h) Pb, Bi, Sb, As, AsO_4^{3-}
4. Borax-bead test (for colored sample) A colorless and transparent loop made on a Pt-wire is touched with a very little sample, heated in oxidizing flame.	4(a) Blue bead (in cold)	4(a) Cu-salt
	(b) Brown bead	(b) Ni or Fe-salt
	(c) Deep blue bead	(c) Co-salt
	(d) Green bead	(d) Cr-salt
	(e) Amethyst-violet bread	(e) Mn-salt
5. Fusion test (for colored sample) 0.1 g sample + Na_2CO_3 (0.3g) + little $NaNO_3$, mix well + 2 beads NaOH, heat by blow pipe in oxidizing flame on a mica-foil, extract the yellow /green mass with water.	5(a) Yellow mass; The extract + dil AcOH + lead acetate soln. → a yellow ppt.	5(a) Cr-salt
	(b) Green mass; The extract + dil AcOH → a pink colored soln.	(b) Mn-salt
6 Test of NH4+: 0.1 g sample + conc. NaOH soln., heated (carefully)	6 Smell of ammonia.	6 NH_4^+- salt

7. Drop test (a) sample + $(NH_4)_2S$ + 2 drops of dH	7(a)(i). Blackening	7(a)(i) Hg, Pb, Bi, Cu, Ni or Co-salt
	(ii) Yellow or orange	(ii) Cd, As, Sb or Sn-salt
(b) conc. NH_4OH.	(b) (i) Blackening	(b)(i) Hg_2^{2+}
	(ii) Intense blue colour	(ii) Cu salt,
	(iii) Light blue	(iii) Ni-salt
8. Fluorescence test 0.1g sample in a 100ml beaker + a little Zn dust (Sn free) + 2ml cH, stir with a tt filled with water, held the tt in oxidizing flame.	8. Blue fluorescence	8. Sn-salt

DRY TESTS FOR ACID RADICALS

EXPERIMENT	OBSERVATION	INFERENCE
1. Sample + dS (2ml), warm (tt)	1. (a) Gas with rotten egg smell, turning $Pb(OAc)_2$ paper black.	1(a) S^{2-} may be present
	(b) Gas, turning acidified $K_2Cr_2O_7$ paper green.	(b) SO_3^{2-}, S^{2-}, $S_2O_3^{2-}$
	(c) Evolution of brown fumes	(c) NO_2^-
	(d) Same as (b) accompanied with turbidity for sulfur.	(d) $S_2O_3^{2-}$
2. Sample + Zn-dust + dS, heat (tt)	2. Gas - turning $Pb(OAc)_2$ paper black	2. S^{2-}, (SCN^- interferes)
3. Iodine-azide (2 drops) on a clean watch glass + sample (pinch).	3. Evolution of N_2 gas	3. SCN^-, S^{2-}, or $S_2O_3^{2-}$ (catalyze $I_2 + 2NaN_3 \rightarrow 2NaI + 3N_2$)
4. (a) Sample + cS, warm.	4. (a) (i) Oily bubbles	4. (a)(i) F^- may be present
	(ii) Violet vapor	(ii) I^-
	(iii) Brown fumes	(iii) Br^-, NO_2^-
(b) Sample + cS + Cu-turnings,	(b) Brown fumes	(b) NO_3^-
5. Sample + PbO_2 + dil. AcOH, warm	5. (a) Brown fumes turning	5. (a) (i) I^- or Br^- present

	(i) Fluorescein-paper red (ii) Schiff's reagent violet	(ii) Br' confirmed
	(b) Violet vapor, dissolve in a drop of KBr soln. and added it to starch soln. → blue color.	(b) I' confirmed
6. Chromyl Chloride test: Solid sample + $K_2Cr_2O_7$ + cS, heat gently.	6. Red vapor, dissolve it in dil NaOH soln → a yellow soln + AcOH + $Pb(OAc)_2$ soln → a yellow ppt.	6. Cl⁻ present
7. Solid sample + dil. NaOH, boil till no smell of NH_3 + pinch of Devarda's alloy, warm.	7. Smell of NH_3	7. NO_2^-, NO_3^-, $Fe(CN)_6^{3-}$ etc. may be present.

TESTS FOR INTERFERING ACID RADICALS

EXPERIMENT	OBSERVATION	INFERENCE
8. (a) 3ml 1:1HNO$_3$ extract of sample + 5ml of $(NH_4)_2MoO_4$ reagent, warm (40°C)	8. (a) A canary yellow ppt.	8. (a) PO_4^{3-} may be present
(b). Same as test 8(a) except boil instead of warm.	(b) Yellow ppt. [NB: brown fumes]	(b). AsO_4^{3-} may be present [NB: AsO_3^{3-} may be present]
9. dN extract of sample + $(NH_4)_2MoO_4$ - tartaric acid reagent, warm.	9. A canary yellow ppt.	9. PO_4^{3-} confirmed
10. Heat a Na_2CO_3 - bead made in Pt-loop with a pinch amount of solid sample in the flame, cool, dissolve in 1:1 HNO$_3$ by boiling + 3ml of $(NH_4)_2MoO_4$ reagent, boil, centrifuge, filter. Filtrate + 2-3 crystals of oxalic acid, warm + 1 drop of benzidine reagent then 2 drops of NaOAc soln (saturated).	10. Blue coloration	10. SiO_3^{3-} confirmed

11. Solid sample + 2ml MeOH + 2 drops cS, boil and ignite the vapors.	11. Green flame on test-tube mouth.	11. Borate confirmed
12. (If test 11 be positive) Solid sample + 2ml MeOH, boil and ignite the issuing vapors.	12. Green flame	12. H_3BO_3 confirmed
13. BF_3 test (in presence of Cu and Ba-salts): 0.1g sample + 0.1g CaF_2 + cS, make a paste and hold by a glass rod near the base of oxidizing flame.	13. Green flame	13. Borate confirmed
14. Solid sample + ($CaSiO_3$) + cS, heat and hold a moist glass rod over the issuing gas.	14. Oily appearance of the tt and the moist glass rod becomes turbid. Wash the turbid drop in a tt with a little water + 2 drops alizarin or alizarin-S reagent + 2 drops zirconyl nitrate reagent → yellow coloration. OR Wash the turbid drop in a tt with a little water + 2 drops	14. F^- confirmed

| | $(NH_4)_2MoO_4$ reagent, warm, cool + 1 drop of benzidine reagent then 2 drops of NaOAc soln (saturated) → a blue coloration. | |

Reactions:

8: $Na_2HPO_4 + 12(NH_4)_2MoO_4 + 23HNO_3 \rightarrow$
$(NH_4)_3[PO_4.12MoO_3] + 2NaNO_3 + 21NH_4NO_3 + 12H_2O$

(ammonium phosphomolybdate, a canary yellow solid)

Similarly, for AsO_4^{3-}, ammonium arsenomolybdate, a yellow solid is formed. AsO_3^{-3} and As^{+3} are oxidized in HNO_3 to AsO_4^{-3} and also respond this test. Benzidine (p, p'-diaminobiphenyl) is oxidized by phosphomolybdic acid or its insoluble ammonium salt and gives a blue oxidation product. AsO_4^{-3} and SiO_3^{-3} react with $(NH_4)_2MoO_4$ to give the ammonium salt of arsenomolybdic acid, $H_3[AsO_4.12MoO_3]$ and silicomolybdic acid, $H_4[SiO_4.12MoO_3]$.

14: $2F^- + H_2SO_4 \rightarrow 2HF + SO_4^{-2}$;

$CaSiO_3 + H_2SO_4 \rightarrow CaSO_4 + SiO_2 + H_2O$; $SiO_2 + 4HF \rightarrow SiF_4 + 2H_2O$;

$3SiF_4 + 4H_2O \rightarrow H_4SiO_4$ (silicic acid) $+ 2H_2SiF_6$. Silicic acid reacts with $(NH_4)_2MoO_4$ to give ammonium silicomolybdate, $(NH_4)_4[SiO_4.12MoO_3]$ which oxidizes benzidine to benzidine blue.

Preparation of Benzidine acetate reagent: 0.05g Benzidine + 10ml glacial AcOH + 90ml water.

WET TESTS FOR ACID RADICAL:

Na$_2$CO$_3$-Extract Preparation:

Sample (0.2g) + Na$_2$CO$_3$ (0.6g) in a conical flask with a funnel + 10ml distilled water, boiled for 20 - 30 min maintaining the volume constant by the addition of water, cooled and filtered. Filtrate is Na$_2$CO$_3$-Extract.

EXPERIMENT	OBSERVATION	INFERENCE
1. 1ml extract + dS (a few drops excess) + 3 drops dil. KMnO$_4$	1. Decolonization of pink color.	1. Reducing agents (S^{2-}, SO$_3^{2-}$, S$_2$O$_3^{2-}$, NO$_2^-$, Br$^-$, AsO$_3^{3-}$ etc.)
2. 1ml extract + cH (a few drops excess) + 3 drops MnCl$_2$ reagent, heat in a water bath for 1-2 min.	2. Brown or black coloration or precipitation.	2. Oxidizing agents (NO$_3^-$, NO$_2^-$, BrO$_3^-$, IO$_3^-$, CrO$_4^{2-}$, Fe(CN)$_6^{3-}$ etc.)
3. 1ml extract + a few drops of sodium nitroprusside soln.	3. Transient violet coloration.	3. S^{2-} (except CdS, PbS, ZnS etc.) confirmed.
4. 1ml extract + dN + 1ml AgNO$_3$ soln.	4(a) Curdy white ppt. readily soluble in NH$_4$OH.	4(a) Cl$^-$, SCN$^-$
	(b) Dirty white ppt. soluble in excess conc. NH$_4$OH.	(b) Br$^-$

	(c) Light yellow ppt. insoluble in conc. NH_4OH.	(c) I^-
	(d) White ppt changing through yellow, brown to black. [NB: very dilute soln responds this test.]	(d) $S_2O_3^{2-}$ confirmed.
5. 1 drop extract + 1 drop of $FeCl_3$ soln.	5(a) Deep blue ppt.	5(a) $Fe(CN)_6^{4-}$ confirmed.
	(b) Deep red coloration.	(b) SCN^- confirmed.
6. 1 drop extract + 1 drop of $FeSO_4$ soln.	6. Deep blue ppt.	6. $Fe(CN)_4^{3-}$ confirmed
7. (a) 1ml extract + dH + 1ml of $BaCl_2$ soln.	7. (a)(i) White ppt.	7. (i) SO_4^{2-}
	(ii) Light yellow ppt. soluble in dN.	(ii) CrO_4^{2-}, $Cr_2O_7^{2-}$ (if the extract is yellow or orange in color)
(b) 1ml of $BaCl_2$ soln. + 1 drop of $KMnO_4$ + 1ml extract + H_2O_2 to discharge the color of $KMnO_4$, centrifuge.	(b) Adsorption of pink color by the white ppt.	(b) SO_4^{2-} confirmed

8. (a) Few drops extract + dil. AcOH + 1 drop of sulphanilic acid + 1 drop of α-naphthylamine reagent, stir.	8. (a) A deep red coloration.	8. (a) NO_2^- confirmed
(b) (If NO_2^- absent) Same as test 8a + Zn dust.	(b) A deep red coloration.	(b) NO_3^- confirmed
(c) (<u>If NO_2^- present</u>) 1ml extract + 0.1g sulphamic acid, or NH_4Cl, boil for a few minutes (to remove nitrite as N_2) and perform test 8b.	(c) A deep red coloration.	(c) NO_3^- confirmed
9. 1 drop extract + starch soln. + Cl_2 water or a crystal of $NaNO_2$.	9. Blue Color	9. I^- confirmed
10(a) 1 ml extract + 2 drops of H_2SO_3 + 1ml CCl_4, shake.	10. (a)(i) yellow organic layer	10. (i) BrO_3^- may be present
	(ii) violet organic layer	(ii) IO_3^- confirmed
(b) 1 drop extract + 1 drop $MnSO_4$	(b) Blue coloration	(b) BrO_3^- confirmed

reagent + 1 drop dS (→ red coloration), warm for 2 min + 1 drop benzidine + 2 drops NaOAc.		
(c) 1 drop soln + 2 drops dH + 1 drop of $(NH_4)_2S$ soln.	(c) Intense blue violet coloration	(c) IO_3^- confirmed
11. 1 drop the soln.+ 2 drops of dH + 1 drop $(NH_4)_2S$ soln.	11 Yellow ppt.	11 Arsenite (AsO_3^{3-}) may be present.
12. 1ml soln. +2 drops magnesia mixture + 2 drops conc. NH_4OH.	12 White ppt.	12 PO_4^{3-} and AsO_4^{3-} may be present
13. (If the soln. be yellow) 1 drop soln. + 2 drops glacial HOAc + 1 drop $Pb(OAc)_2$ in a watch glass.	13 Yellow ppt.	13 CrO_4^{2-} confirmed

SPECIAL TESTS FOR SIMILAR ACID RADICALS

1. Cl⁻, Br⁻, I⁻, SCN⁻, $S_2O_3^{2-}$

EXPERIMENT	OBSERVATION	INFERENCE
1. OS + PbO_2 + dil AcOH, warm (T)	1. (a) Brown fumes turning (i) Fluorescein-paper red (ii) Schiff's reagent violet	1. (a) (i) I' or Br' present (ii) Br' confirmed
	(b) Violet vapor, dissolve in a drop of KBr soln. and added it to starch soln. → blue color.	(b) I' confirmed
2. 1 drop Na_2CO_3 extract + starch soln. + Cl_2 water or a crystal of $NaNO_2$.	2. Blue Color	2. I⁻ confirmed
3. 1 drop Na_2CO_3 extract + 1 drop of $FeCl_3$ soln.	3. (a) Deep red coloration.	3. (a) SCN⁻ confirmed.
4. 1ml Na_2CO_3 extract + dN + 1ml $AgNO_3$ soln.	4. (a) Curdy white ppt. readily soluble in NH_4OH.	4. (a) Cl⁻, SCN⁻
	(b) Dirty white ppt. soluble in excess conc. NH_4OH.	(b) Br⁻

	(c) Light yellow ppt. insoluble in conc.NH₄OH.	(c) I^-
	(d) White ppt. changing through yellow, brown to black. [*NB: very dilute soln. responds this test.*]	(d) $S_2O_3^{2-}$ confirmed.
5. The Na₂CO₃ extract is boiled with excess AcOH and a pinch of PbO₂, when all the radicals except Cl⁻, are decomposed. The perfectly colorless soln. having no trace of Br₂ or I₂ is acidified with 3 drops of dN, boiled , added 2 drops of dN, boiled and then 2 drops of AgNO₃ is added	5. white ppt. of AgCl soluble dil. NH₄OH	5. Cl⁻ confirmed

2. Br^-, I^-, and IO_3^-:

To the soln. (3ml), a few drops of $BaCl_2$ soln. are added and filtered.	
Residue: Washed with 1ml portion (twice) of water, dissolved ppt. in dH. + 3 drops dS, centrifuged	Filtrate: 1 ml of CCl_4 added followed by a few drops of Cl_2 water, shaken → a violet color confirms I^-. More Cl_2 water is added till the violet layer disappears → brown or yellow color confirms Br^-.
Residue: Rejected	Filtrate: Test for iodate by test no. by 10(c).	

3. SCN^-, $Fe(CN)_6^{-3}$, $Fe(CN)_6^{-4}$

EXPERIMENT	OBSERVATION	INFERENCE
To a filter paper impregnated with a drop of $FeCl_3$, one drop the test soln. then one drop of dH and finally 2 drops of water is added. After waiting some time, the following concentric rings will be observed:	(a) Interior blue spot	$Fe(CN)_6^{4-}$
	(b) Green circle	$Fe(CN)_6^{3-}$
	(c) Outer red ring	SCN^-

4. S^{-2}, SO_3^{-2}, SO_4^{-2}, and $S_2O_3^{-2}$

Na$_2$CO$_3$ extract (5ml) + CdCO$_3$ (excess), shake, filter			
Residue: Yellow CdS (CdCO$_3$, dissolve in AcOH) **S^{-2} confirmed**	Filtrate: + Sr(NO$_3$)$_2$ soln. (slight excess), shake, filter		
	Residue: + dH, filter		Filtrate: + 3 drops AgNO$_3$ soln. → white ppt. changing through yellow, brown to black. **$S_2O_3^{-2}$ confirmed**
	Residue: White (SrSO$_4$), insoluble in any acid. **SO_4^{-2} confirmed**	Filtrate + 2 drops I$_2$ soln. → discharge of I$_2$-color. **SO_3^{-2} confirmed**	

5. PO_4^{-3}, AsO_4^{-3} and AsO_3^{-3}

Na$_2$CO$_3$ extract (5ml) + 3ml magnesia mixture, shake, filter	
Residue: White (PO_4^{-3}: AsO_4^{-3}) + dH, boil to dissolve + NaHSO$_3$ (to reduce AsO_4^{-3} to AsO_3^{-3}) + H$_2$S, filter		Filtrate: + dH, warm + H$_2$S→Yellow ppt. (As$_2$S$_3$) **AsO_3^{-3} confirmed**
Residue: Yellow (As$_2$S$_3$) **AsO_4^{-3} confirmed**	Filtrate: Boil off excess H$_2$S + (NH$_4$)$_2$MoO$_4$ reagent, warm → Canary yellow ppt. **PO_4^{-3} confirmed**	

PRELIMINARY WET TESTS FOR BASIC RADICALS:

ANALYSIS OF cN-EXTRACT:

The sample was boiled with cN (5ml) till the evolution of brown fumes ceased; diluted with water (5ml), centrifuged if reqd., and with the supernatant the following tests were performed.

EXPERIMENT	OBSERVATION	INFERENCE
1. 1 ml of the extract + 2 drops of dS.	1. White ppt.	1. Ba, Sr or Pb may be present
2. 2 ml extract + pinch of $NaBiO_3$ in cold.	2. Pink coloration.	2. Mn confirmed
3. 1 drop extract + 1 drop of $K_4Fe(CN)_6$ soln.	3. Blue ppt.	3. Fe confirmed
4. 1 drop extract+2 drops of conc.NH_3 (make ammonical) + 1 drop of dimethylglyoxime reagent + a few drops of tartatric acid (5%)	4(a) Red ppt.	4(a) Ni confirmed
	(b) Soln. turned blue after adding NH_3	(b) Cu present

ANALYSIS OF KOH EXTRACT:

About 100 mg of the sample was boiled for 5 min. with 10 ml. of water. And 8 pellets of KOH in a beaker covered with a clock glass; cooled, centrifuged and with the supernatant the following tests were performed.

EXPERIMENT	OBSERVATION	INFERENCE
1. 1 ml. of soln. was acidified with dH, if any ppt., the whole soln. should be acidified, filtered and the following tests should be performed with the filtrate.	1(a) White ppt. soluble in boiling water, to the soln. KI soln. added, boiled and cooled. → Golden Spangles	1(a). Pb confirmed
	(b) Orange ppt.	(b) Sb, S^{-2} may be present
	(c) Yellow ppt.	(c) As, Sn, S^{-2} may be present
2. To the clear alkaline soln. a drop of Alizarin-S reagent + HOAc till color changes + 1 drop excess HOAc.	2. Red ppt. or color"	2. Al confirmed
3. 1 drop extract +2 drops dS +1 drop 0.1% $CuSO_4$ + 1 drop of $(NH_4)_2Hg(CNS)_4$ reagent [mixture of equal volume of NH_4SCN and $HgCl_2$]	3. Violet ppt.	3. Zn confirmed

4. To 5ml extract, 5ml of cH added and (a) To 2ml of the soln, 2ml of $SnCl_2$ added, warmed in a water bath.	4(a) Blackening	4(a). As confirmed
(b) To 2ml soln.,a pinch of $NaNO_2$ added, warmed, cooled, 2 drops was added to 5 drops Rhodamine B reagent.	(b) Blue color	(b) Sb confirmed
(c) 2ml soln. was warmed for 1 min. with Mg-turning (water bath), cooled, to 1 drop of it was added 1 drop of Cacothelein reagent.	(c) Red or Violet coloration	(c) Sn confirmed

WET TESTS FOR BASIC RADICALS

PREPARATION OF SOLUTION:

Boil the sample (200 mg) for one min. with 8 ml of water; filter:	
Residue: Boil with dH (3ml). If any residue, add 3 ml cH and boil for 1 min., filter.	Filtrate: **Soln.(A)**: Reject if there is no residue on evaporation (**aq. soln.**)
Residue: Boil with 4ml aqua regia (1ml cN + 3ml cH), evaporate to dryness, extract with 3ml dH and filter.	Filtrate: **Soln. (B)** treated as in **HCl soln.**
Residue: **Insoluble part**	Filtrate: **Aqua regia soln.**
NB: Mix **HCl soln.** and **Aqua regia soln.** and perform group analysis with this **acid soln.** (C)	

Tips: Mix **aq. Soln.** (after Gr-I separation) with **HCl soln.** and **aqua regia soln.** and perform group analysis. On mixing if any ppt. [e.g., SO_4^{2-} (in aq. soln.) + Ba^{2+} (in acid soln.) → $BaSO4 \downarrow$], filter and then do group analysis.

GROUP ANALYSIS FOR BASIC RADICALS

5ml of soln. + 5 drops dH , filter (not reqd. in C sol.)					
Residue: **Gr-I**: White: Ag, Pb, Hg_2^{+2}	Filtrate: + cH to maintain 0.3N acidity, boil + pass H_2S immediately for 1 min. till complete precipitation and centrifuge.				
	Residue: **Gr-II**: Black: Hg^{+2}, Bi^{+3} Cu^{+2} Pb^{+2} Yellow: Cd^{+2}, As^{+3} $Sn^{+2\ or\ 4}$ Orange: Sb^{+3} ---------- -*Tips*: reduce acidity of the Su after Gr.II separation by NH_4OH & then pass H_2S to precipitate yellow CdS. ---------- -	Filtrate: Boil off H_2S + few drops cN (interfering acid radicals, if any must be removed) + NH_4Cl + NH_4OH till ammoniacal and centrifuge.			
		Residue: **Gr-IIIA**: White: Al^{+3}, Brown : Fe^{+3}, Green: Cr^{+3}	Filtrate: + 2ml NH_4OH, heat + H_2S and centrifuge.		
			Residue: **Gr-IIIB**: Black: Ni^{+2}, Co^{+2}, Pink: Mn^{+2} White: Zn^{+2}	Filtrate: Reduce volume to $1/4^{th}$ by evaporation + NH_4OH + saturated $(NH_4)_2CO_3$ soln. and centrifuge.	
				Residue: **Gr-IV**: White: Ba^{+2}, Sr^{+2}, Ca^{+2}	Filtrate: **Gr-V**: Mg^{+2} Na^+ K^+

Filtrate: **Gr-V**:

(i) 1st part of Filtrate: + NH_4Cl + NH_4OH + Na_2HPO_4 soln., rub inner wall of the tt. with a glass rod → White crystalline ppt. → Mg^{+2} present.

(ii) 2 drops Filtrate + 2 drops **Titan yellow** + 2 drops 0.1(N) NaOH, stir with a glass rod → Red pptn. or coloration → Mg^{+2} present.

(iii) 4 drops Filtrate + 8 drops zinc-uranyl acetate reagent, rub with a glass rod → Yellow ppt. → Na^+ present.

(b) 5 drops Filtrate + conc. NaOH, boil to remove NH_3 completely + AcOH + sodium cobaltinitrite soln. or a pinch of solid, rub with a glass rod → Yellow ppt. → K^+ present.

SEPARATION OF BORATES, F⁻ AND SILICATES: [After Gr-II Separation]

Evaporate the filtrate with cH (3-4 times) to almost dryness, cool + 10ml distilled water, filter and the filtrate is used for Gr-IIIA – V analysis.

PHOSPHATE SEPARATION:

(A) <u>$FeCl_3$-Acetate Buffer Method</u>:

N.B. (1) If phosphate is present and sample is completely soluble in water, do not perform Phosphate separation.

(2) If sample is partly soluble in dil. HCl or completely soluble in dil. HCl and phosphate is present (perform test of phosphate with this solution), then perform phosphate separation.

(i) <u>Test for Fe</u>: 1 ml solution (after Gr II separation) + 2 drops of cN, heat, cool + 2 drops NH_4SCN soln. → thick blood red coloration → **Fe present**.

10 ml soln. (free from Gr I & II cations)			
a) + 2g NH$_4$Cl, heat to dissolve, cool + NH$_4$OH dropwise till a faint permanent ppt. appears.			
b) + 10 ml acetate-buffer (pH 4.6). Two cases may arise:			
(1) if any ppt. → Gr-IIIA present.			
(2) If no ppt. → Gr-IIIA absent.			
[don't filter any ppt. in this stage]			
c) + Neutral FeCl$_3$ soln. dropwise with stirring until the soln. acquires tea color.			
d) Dilute two times the mixture with water, boil (2min), filter hot, wash the residue with little boiling water.			
Residue: [Gr-IIIA present, case b (1)] (i) Transfer the ppt. into a tt by 5ml. water (ii) + 5 ml NaOH + 5 ml 3% H$_2$O$_2$ (iii) Boil (2min) filter & wash with water		Filtrate: [Gr-IIIA absent case b(2)] Boil to reduce the volume to 5ml + 0.5g NH$_4$Cl + NH$_4$OH, (filter off any ppt.), warm + H$_2$S, filter	
Residue [Reject]	Filtrate (yellow Na$_2$CrO$_4$)	Residue **Gr-IIIB present**	Filtrate **Gr-IV & V**
	1st part of the Soln. +1g NH$_4$Cl, boil → white ppt. of Al(OH)$_3$ **Al^{+3} present**	2nd part of the + AcOH + Pb(OAc)$_2$ soln. → yellow ppt. of PbCrO$_4$ **Cr^{+3} present**		

(B) Zirconyl nitrate method: [(ZrO(NO$_3$)$_2$]

10ml soln. (free from Gr I & II cations) Boil off H$_2$S, adjust acidity to 1(N)+ 0.5g NH$_4$Cl, stir to dissolve + zirconyl nitrate soln. (dropwise) until precipitation is completed, heat to boiling with stirring, filter and wash with a little hot water.		
Residue: Zr$_3$(PO$_4$)$_4$ - reject	Filtrate: (If no pptn. on addition of 1 drop Zr(NO$_3$)$_4$, phosphate is completely removed) + 0.5g NH$_4$Cl, heat + NH$_4$OH, boil for 2 min and filter.	
	Residue: Gr-IIIA + excess Zr^{+4}	Filtrate: Gr-IIIB, IV, V

ANALYSIS OF INDIVIDUAL GROUP PRECIPITATE

(A). ANALYSIS OF Gr-I PRECIPITATE

Gr-I ppt. (AgCl, Hg_2Cl_2, $PbCl_2$) + water, boil, filter in hot condition				
Residue: (White: AgCl, Hg_2Cl_2) + 5ml NH_4OH over the ppt. on the filter paper		Filtrate: (Hot soln. of $PbCl_2$)	
Residue: **Black** **[Hg + $Hg(NH_2)Cl$]** **Hg_2^{+2} present**	Filtrate: + dN **Curdy white ppt.** of AgCl **Ag^+ present**	1st part + K_2CrO_4 soln. **Yellow ppt.** of $PbCrO_4$	2nd part + KI soln. **Yellow ppt.** of PbI_2, soluble on heating, reappears as golden spangles on cooling
		\multicolumn{2}{c}{**Pb^{+2} present**}	

(B). ANALYSIS OF Gr.II PRECIPITATE

Gr-II ppt. (HgS, PbS, Bi_2S_3, CuS, CdS, As_2S_3, Sb_2S_3, SnS, SnS_2) + 5ml 2(N) KOH soln., boil for 3min + H_2S-water, stir and filter.	
Residue: Gr-IIA (HgS, PbS, HgS, PbS)	Filtrate: Gr-IIB (thiosalts of As, Sb, Sn)

Gr-IIA

Gr-IIA (HgS, PbS, HgS, PbS) + 5ml dH(1:3), boil, filter, wash with a little water			
Residue: Black (HgS) Hg^{+2} confirmed	Filtrate: (if white ppt. on adding 2ml mixture of dS & EtOH into1ml filtrate + dS (1:1), heat, filter		
	Residue: White ($PbSO_4$) Pb^{+2} confirmed	Filtrate+ NH_4OH, filter:	
		Residue: White [$Bi(OH)_3$] Bi^{+3} confirmed	Filtrate: (i) 1st part (if blue) + $K_4Fe(CN)_6$ → reddish brown ppt. Cu^{+2} confirmed (ii) 2nd part + conc. HCl, warm+ H_2S, filter off CuS, reduce acidity by NH_4OH, warm + H_2S→ yellow ppt. Cd^{+2} confirmed

Gr-IIB

Gr-IIB (thiosalts of As, Sb, Sn) + acidify with cH + H_2S filter and wash the ppt.		
Residue: (As_2S_3, Sb_2S_3, SnS, SnS_2) + 5ml cH+5ml water, boil and filter		Filtrate: reject
Residue: Yellow (As_2S_3) As^{+3} confirmed	Filtrate:	
	1st part, make ammoniacal just 2g oxalic acid+H_2S→Orange ppt. (Sb_2S_3) Sb^{+3} confirmed	(i)2nd part + Fe nails, heat, filter into $HgCl_2$ soln. →White ppt. (Hg_2Cl_2) Sn^{+2} confirmed (ii) 3rd part + Al powder, heat, filter. powder, heat, filter. +2 drops Cacotheline soln. → violet coloration Sn^{+2} confirmed

(C). ANALYSIS OF Gr-IIIA PRECIPITATE

Gr-IIIA ppt. [Fe(OH)$_3$, Cr(OH)$_3$, Al(OH)$_3$] Transfer in a tt + 5ml water + 5ml NaOH + 5ml 3% H$_2$O$_2$, boil, filter, wash with water	
Residue: Reddish brown [Fe(OH)$_3$] + dH to dissolve + K$_4$Fe(CN)$_6$ → Deep blue ppt. **Fe^{+3} confirmed**	Filtrate: (i) 1st part of the soln. +1g NH$_4$Cl, boil → white ppt [Al(OH)$_3$] **Al^{+3} confirmed** (ii) 2nd part of the soln. (if yellow) + AcOH + Pb(OAc)$_2$ soln. →yellow ppt. (PbCrO$_4$) **Cr^{+3} confirmed**

(D). ANALYSIS OF Gr-IIIB PRECIPITATE

Gr-IIIB ppt. (NiS, CoS, MnS, ZnS) + dH, shake, allow to stand for 2min, filter			
Residue: Black (NiS, CoS) + aqua regia to dissolve ppt., evaporate to dryness, extract with water		Filtrate: boil off H$_2$S + NaOH soln, boil, cool, filter	
(i) 1ml ext. + NH$_4$Cl+ NH$_4$OH + 5 drops DMG reagent → rose red ppt **Ni^{+2} confirmed**	(ii) 1ml ext. +1ml amyl alc. + 1ml +cH+1g NH$_4$SCN, shake well→ blue org. layer **Co^{+2} confirmed**	Residue: Brown (MnO$_2$xH$_2$O) +HNO$_3$ (1:1) to dissolve + pinch of NaBiO$_3$, allow to stand, violet coloration **Mn^{+2} confirmed**	Filtrate: (i) 1st part + AcOH + H$_2$S→white ppt. (ZnS) **Zn^{+2} confirmed** (ii) 2nd part + dS +3drop CuSO$_4$/Co(NO$_3$)$_2$ + NH$_4$- mercuric thiocyanate, shake for 5min→ a light blue ppt. **Zn^{+2} confirmed**

(E). ANALYSIS OF Gr-IV PRECIPITATE

Gr-IV ppt. ($BaCO_3$, $SrCO_3$, $CaCO_3$) + dil. AcOH, heat to dissolve + K_2CrO_4 soln. dropwise, filter hot			
Residue: Yellow ($BaCrO_4$) **Ba^{+2} confirmed**	Filtrate: + NH_4OH +$(NH_4)_2CO_3$ soln. (excess), filter, wash with hot water			
	Residue: White ($SrCO_3$, $CaCO_3$) + dil. AcOH, heat to dissolve, boil off CO_2 + saturated $(NH_4)_2SO_4$ soln., heat on a water bath for 2-3min, filter		Filtrate: Reject
	Residue: White ($SrSO_4$) **Sr^{+2} confirmed**	Filtrate: + NH_4-oxalate soln., heat to boiling White ppt. giving transient brick red flame **Ca^{+2} confirmed**	

ANALYSIS OF INSOLUBLE PART

[SiO_2, Al_2O_3, SnO_2, CaF_2 (white), Fe_2O_3 (ig., brown), Cr_2O_3 (green), $PbCrO_4$ (fused, yellow / brown), $PbSO_4$, $SrSO_4$, $BaSO_4$, various silicate (white)]

EXPERIMENT	OBSERVATION	INFERENCE
1. **Test for $PbSO_4$ (white)**: Insoluble part + conc. ammonium acetate soln., heat a) 1-part soln. + CH_3COOH + K_2CrO_4 soln. b) 1-part soln. + dil. HCl + $BaCl_2$ soln.	1. Soluble a) yellow ppt. b) white	1. $PbSO_4$ present a) Pb^{+2} present and confirmed b) SO_4^{-2} present and confirmed
2. **Test for $PbCrO_4$ (yellow)**: Insoluble part + NaOH (3 beads) + 2ml water, boil, acidify with dil. HCl a) Cool the test tube, filter. The yellow filtrate + dil. AcOH + $Pb(OAc)_2$ soln.	2. White ppt. which dissolves on heating and reappears on cooling. a) yellow ppt.	2. Pb^{+2} present and confirmed a) CrO_4^{-2} present and confirmed
3. **Test for Al_2O_3, SiO_2, SnO_2 (White)**: Insoluble part + fusion mixture (3 times) + NaOH (3 beads), fuse on a nickel spoon,	3. a) (i) White gelatinous ppt. (perform alizarin-S test)	3. a) (i) Al^{+3} present and confirmed

extract with hot water and filter. a) (i) 2 ml filtrate + NH$_4$Cl (lg), boil. (ii) 2ml filtrate + conc. HCl (iii) 2 ml filtrate + conc. HCl + H$_2$S b) <u>Residues</u>: Carbonates of Ca^{+2}, Ba^{+2}, Sr^{+2} (perform flame test). Proceed as Gr IV ppt.	(ii) Sand like white ppt., does not dissolve on boiling. (iii) Yellow ppt. (SnS$_2$)	(ii) SiO$_2$ present and confirmed. (iii) Sn^{+4} present and confirmed
4. <u>Test for Cr$_2$O$_3$ (green)</u>: Green insoluble part + Na$_2$O$_2$, fuse, extract with water. The yellow extract + dil. AcOH + Pb(OAc)$_2$ soln.	4. Yellow ppt.	4. Cr$_2$O$_3$ present and confirmed
5. <u>Test for Fe$_2$O$_3$ (brown)</u>: Insoluble part + KHSO$_4$, fuse, extract with dil. H$_2$SO$_4$. 2ml extract + K$_4$Fe(CN)$_6$ soln.	5. Prussian blue ppt.	5. Fe$_2$O$_3$ present and confirmed

4
Special Notes on Mixtures of Ions

Aqua regia extract should (if aqua regia soluble part presents) be evaporated to dryness and the residue is to be analyzed after dissolving it in water or dilute HCl because HNO_3 being an oxidizing agent, reacts with H_2S (during group analysis) to give much colloidal sulfur.

Salts which are soluble in a particular solvent (water, dilute HCl) solvent may react among themselves on solution by double decomposition to produce insoluble salts.
  e.g. (a) Both Na_2SO_4 and $BaCl_2$ are soluble in dil HCl separately but their mixture in dilute HCl interact to give insoluble $BaSO_4$.
  $Na_2SO_4 + BaCl_2 \rightarrow BaSO_4\downarrow + 2NaCl$
  (b) A mixture of As_2O_3 and ZnS on treatment with dilute HCl gives yellow precipitate of As_2S_3.

If the sample completely dissolves in HCl and on dilution or cooling does not give precipitate, Gr-I metal ions are absent (except Pb^{2+}).

Dilute HCl solution of sample may contain Pb^{2+}. Upon cooling or addition of concentrated HCl Pb^{2+} will be precipitated as $PbCl_2$, however, some portion of Pb^{2+}

remains in the solution and it precipitated as black PbS in Gr-II on passing of H_2S in acidic solution.

Addition of HCl to aqueous solution of the sample may precipitate Boric acid and Silicic acid partially.

HCl solution of the sample on dilution with water, turns milky due to the precipitation of Sb, Bi & Sn. Milkiness disappears on addition of HCl.
$$BiCl_3 + H_2O \rightarrow BiOCl + 2HCl$$

Brown color of the Group-IIIB filtrate indicates the presence of colloidal NiS (Produced by passage of H_2S for a long time). It may be coagulated by boiling with acetic acid and then tested for Ni^{2+} directly.

The filtrate after Gr-IIIB usually contains much NH_4^+. Acidic NH_4^+ reduce the concentration of CO_3^{2-} upon addition of $(NH_4)_2CO_3$.
$$NH_4^+ + CO_3^{2-} \rightarrow NH_3 + HCO_3^-$$
Lower the concentration of CO_3^{2-} may lead to incomplete precipitation of Gr-IV metal carbonates. So excess NH_4^+ should be removed first by adding concentrated HNO_3.
$$NH_4Cl + HNO_3 \rightarrow N_2O + HCl + 2H_2O$$

Alkaline earth metals' bicarbonates are soluble, so while precipitating Gr-IV metal carbonates, the solution should be kept at 60 °C to decompose any bicarbonate (present as impurity or formed in reaction medium) but the solution should not be boiled as the reaction is reversible.
$$BaCO_3 + NH_4Cl \rightarrow BaCl_2 + (NH_4)_2CO_3$$

Boiling decomposes $(NH_4)_2CO_3$, so the equilibrium shifts right, thus the metal carbonates dissolve.

Phosphates of metals of Groups-IIIA, IIIB, IV and Mg, also, borates and fluoride of Groups IIIB, IV and Mg are insoluble in ammonia medium. So, they may precipitate on Group-IIIA. For example, silicates may precipitate as silicic acid and this may lead to confusion with precipitate of $Al(OH)_3$. Thus, these interfering acid radicals must be removed before the precipitation of Gr-IIIA.

Before precipitating metal ions of Group-IIIA, concentrated HNO_3 should be added to oxidize Fe^{2+} to Fe^{3+}, otherwise, Fe^{2+} will precipitate incompletely as $Fe(OH)_2$ by NH_4OH and NH_4Cl.

Phosphate separation becomes necessary only when there will be precipitation on addition of NH_4Cl and NH_4OH to the solution after Gr-II. In case of absence of Gr-IIIA, IIIB, IV & Mg^{2+}, No precipitate will appear on addition of NH_4OH and NH_4Cl.

Ba^{2+}, Sr^{2+}, Ca^{2+}, tend to precipitate during the separation of phosphate by $FeCl_3$ method. So it is suggested to test for their presence before the separation of phosphate.

1. To a solution after Gr-II containing ammonium acetate and acetic acid buffer, a few drops of K_2CrO_4 solution is added. → yellow precipitate of $BaCrO_4$ → Ba^{2+} is present.

2. The yellow ppt. is filtered off and the resulted filtrate is boiled with $(NH_4)_2SO_4$ solution → white

precipitate of SrSO$_4$, does not dissolve in any acid. → Sr^{2+} is present.

3. 1 ml of solution containing buffer is boiled, then a few drops of ammonium oxalate solution is added and shaken → white ppt. of CaC$_2$O$_4$ → Ca^{2+} is present.

Oxidizing and Reducing anions.

Oxidizing anions	NO$_3^-$, NO$_2^-$, CrO$_4^{2-}$, MnO$_4^-$, XO$_3^-$ (X = Cl, I, Br), Fe(CN)$_6^{3-}$, AsO$_3^{3-}$.
Reducing Anions	NO$_2^-$, Br$^-$, I$^-$, CN$^-$, SO$_3^{2-}$, S^{2-}, S$_2$O$_3^{2-}$, AsO$_3^{3-}$, Fe(CN)$_6^{4-}$.

The yellow precipitate in Gr-II refers to the presence of Cd^{2+} or As^{3+} precipitated as CdS or As$_2$S$_3$. But CdS is insoluble in NaOH/KOH solution, whereas As$_2$S$_3$ is soluble in hot NaOH/KOH solution.

In Gr-II, for the precipitation of CdS, the solution must be 0.3 (N) HCl, dilute or concentrated HCl solution does not precipitate CdS during the H$_2$S(g) pass.

5
Difficulties and Their Removals During Group Analysis

Presence of $S_2O_3^{2-}$
Difficulties:
Addition of HCl in Gr-I, colloidal sulfur precipitates. So, it becomes difficult to filter.

Removal:
Add some NH4Cl + small pieces of filter paper → boil → filter. Filtrate is treated as Gr-II.

Presence of S^{2-}
Difficulties:
A yellow, orange or black ppt. may arise on addition of HCl in Gr-I.

Removal:
Filter the residue, Residue is treated as Gr-II precipitate. In the filtrate, pass H₂S if any ppt. occur, filter → residue is Gr-II ppt.

Filtrate is treated as Gr-IIIA onwards.

Presence of $[Fe(CN)_6]^{3-}$ and $[Fe(CN)_6]^{4-}$
Removal:
The aqueous solution is treated with 2 ml conc. H₂SO₄ → boil freely → a paste is formed → heat strongly to drive out

acid fumes → add water → boil → treat with HCl and perform group analysis.

Oxidizing agents: Fe^{3+}, MnO_2, $Cr_2O_7^{2-}$, CrO_4^{2-}, AsO_4^{3-} etc.

Difficulties:

When H_2S gas is passed in Gr-II analysis, colloidal sulfur may form.

Removal:

The solution is heated with SO_2 water and oxidizing agents are reduced.

Note: (a) MSO_4 is precipitated if metal ions Ca^{2+}, Pb^{2+}, Ba^{2+}, Sr^{2+} present. Perform prolonged flame test with the residue.

(b) Reaction must be done before passing H_2S in Gr-II. Boil off the SO_2 gas before passing H_2S otherwise the reaction will occur with the remaining SO_2 in solution.

$$SO_2 + H_2S \rightarrow 3S\downarrow + 2H_2O$$

Removal of BO_3^{3-}, F^- and SiO_3^{2-}

The HCl solution after Gr-II is boiled freely to make a paste.

Add 5 ml HCl (conc.) and 5 ml MeOH and boil in low flame → add HCl and water → filter → filtrate is treated as Gr-IIIA.

6
Addition of Some Common Reagents in Water Extract

Remember, the following reactions do not provide the confirmation of the presence of said ions. But you can get some indication of their presence during the experiments.

A. Water extract of anions + AgNO₃ solution

White precipitate	Maybe, CO_3^{2-}, SO_3^{2-}, Cl^-, SCN^-, IO_3^-, BrO_3^-, $Fe(CN)_6^{4-}$, Br^-, I^-
	(a) Precipitate dissolves on addition of excess water extract. $\rightarrow SO_3^{2-}$, $S_2O_3^{2-}$
	(b) Precipitate turns brown on standing/boiling, ppt. dissolves in dil. NH_4OH and HNO_3 $\rightarrow CO_3^{2-}$, BO_2^-, $B_4O_7^{2-}$.
	(c) Precipitate is insoluble in dil. HNO_3 but soluble in dil. NH_4OH $\rightarrow Cl^-$, SCN^-, IO_3^-, BrO_3^-
	(d) Precipitate is insoluble in dil. NH_4OH and HNO_3. Turns orange on warming with conc. HNO_3. $\rightarrow Fe(CN)_6^{4-}$
White ppt. changes color.	With excess $AgNO_3$ reagent, the ppt. rapidly changes color and finally becomes black on standing. $\rightarrow S_2O_3^{2-}$
Black ppt.	Soluble in hot dil. HNO_3 $\rightarrow S^{2-}$
Yellow ppt.	The ppt. dissolves in dil. NH_4OH and HNO_3 $\rightarrow PO_4^{3-}$
Very Pale yellow ppt.	Insoluble in HNO_3 but soluble in concentrated NH_4OH. $\rightarrow Br^-$

Pale Yellow ppt.	Insoluble in strong NH_4OH → I^-
Brownish red ppt.	Soluble in dil. NH_4OH and HNO_3, insoluble in AcOH → AsO_4^{3-}
Yellow ppt.	Soluble in dil. NH_4OH and HNO_3 → AsO_3^{3-}
Brown red ppt.	Soluble in dil. NH_4OH and insoluble in AcOH. → CrO_4^{2-}
Orange red ppt.	Soluble in dil. NH_4OH and insoluble in HNO_3 → $Fe(CN)_6^{3-}$

B. Water extract of anions + $BaCl_2$ solution

White ppt.	(a) Insoluble in dil. HNO_3 → SO_4^{2-}
	(b) Add dil. HNO_3 (i) ppt. dissolves with evolution of odorless gas → CO_3^{2-}. (ii) Gas with smell of burnt sulfur → SO_3^{2-}
	(c) The ppt. is treated with SO_2 water → add CCl_4 → shake → violet organic layer → IO_3^-
	(d) Colloidal sulfur from concentrated solution. → $S_2O_3^{2-}$
Yellow ppt.	Soluble in dil. HNO_3 but insoluble in dil. acetic acid. → CrO_4^{2-}

C. Water extract of anions + $Pb(OAc)_2$ solution

White ppt.	(a) Insoluble in dil. HNO_3 or HCl → SO_4^{2-}
	(b) Evolution of colorless and odorless gas on addition of HNO_3 → CO_3^{2-}
	(c) Evolution of SO_2 gas on addition of dil. HNO_3 → solution turns turbid on standing → SO_3^{2-}
	(d) Soluble ppt. in excess of water extract. The ppt. turns black on boiling → $S_2O_3^{2-}$

	(e) SCN-, Cl-, Br- give hot water soluble ppt. from concentrated solution.
Black ppt.	S^{2-}
Yellow ppt.	(a) dissolves on heating, reappears on cooling → I^-
	(b) Does not dissolve on heating → CrO_4^{2-}

D. Water extract + aq. FeCl₃ solution

Yellowish white ppt.	Insoluble in dil. Acetic acid → PO_4^{3-}
Reddish ppt.	CO_3^{2-}
Dark violet color	The color discharge quickly → $S_2O_3^{2-}$
Dark red color	The color discharge on adding NaF or $HgCl_2$ → SCN^-
Deep blue ppt.	$Fe(CN)_6^{4-}$
Deep brown color	Add CCl_4, shake → violet layer → I^-
Brown color	$Fe(CN)_6^{3-}$

E. Water extract + Hg(NO₃)₂ solution

White ppt.	(a) Soluble in excess of the water extract → SCN^-
	(b) Insoluble in excess of the extract → IO_3^-
Scarlet red ppt.	Soluble in excess of the extract → I^-
Black ppt.	S^{2-}

F. Water extract + KI solution

Extract + KI solution + dil. HCl + CCl_4 → shake	Organic layer turns violet → presence of oxidizing agent/s.
Yellow ppt.	Ag^+, Pb^{2+}, Ba^{2+}
Rose red ppt.	Disappear on addition of excess KI → Hg^{2+}

| Green ppt. | Hg_2^{2+} |
| Yellow coloration | Sb^{3+} |

7
Detection of Rare Elements

Test for Ti^{4+}

Solution in H_2SO_4 (2 drops) + H_2O_2 → Orange yellow color → color discharged by NaF.

Tests for V^{5+}

1. Solution in dil. HCl + H_2S → blue coloration with separation of sulfur.
2. Solution + dil. H_2SO_4 + $FeSO_4$ → blue coloration
3. Solution + conc. HNO_3 + $(NH_4)_2MoO_4$ → Intense yellow coloration due to formation of ammonium molybdo vanadate.
4. Solution + dil. H_2SO_4 + H_2O_2 (2-3 drops) → orange coloration → not discharged by NaF (avoid excess H_2O_2, which will produce yellow coloration).

Tests for W^{6+}

Solution + conc. HCl → light yellow ppt. → boil with $SnCl_2$ → Solution becomes blue.

Tests for Mo^{6+}

Solution + $SnCl_2$ + NH_4SCN → Blood red coloration → discharged by conc. HCl.

Tests for U^{6+}

Solution + dil. HCl + $K_4[Fe(CN)_6]$ → brown ppt. → boiling with NaOH gives a yellow residue of $Na_2U_2O_7$.

Tests for Ce^{4+}

1. Solution + NH_4OH + H_2O_2 → brown ppt.
2. Solution + Oxalic acid (excess) → white ppt. after reduction of Ce^{4+} (the yellow color is discharged before precipitation).

Tests for Zr^{4+}

1. Solution + NaOH + H_2O_2 → white ppt. of ZrO_2.
2. A drop of test solution is taken on a filter paper + 2 deops of p-dimethyl amino benzene azophenyl arsenic acid → red ppt. (not discharged by conc. HCl) → dry the ppt. at low flame → Immerse the filter paper in 2N HCl and warm up to 60 – 70 °C. The deep brown ppt. persists and the excess reagent is washed out.

Tests for Ce^{3+}

Ce^{3+} gives same tests (1 & 2) as for Ce^{4+}.

3. Solution + conc. HNO_3 → yellow coloration.

Bibliography

1. Svehla, G., 2008. *Vogel's Qualitative Inorganic Analysis, 7/e*. Pearson Education India.
2. Moeller, T., 2012. *Chemistry: with inorganic qualitative analysis*. Elsevier.
3. Kolthoff, I.M. and Elving, P.J. eds., 1993. *Treatise on analytical chemistry* (Vol. 13). John Wiley & Sons.
4. Harvey, D., 2000. *Modern analytical chemistry*. Boston: McGraw-Hill Companies, Inc.
5. Charlot, G., 1954. *Qualitative inorganic analysis*. CUP Archive.
6. Feigl, F. and Anger, V., 2012. *Spot tests in inorganic analysis*. Elsevier.

www.ingramcontent.com/pod-product-compliance
Lightning Source LLC
Chambersburg PA
CBHW030010190526
45157CB00014B/2128